The Atom

By Forester de Santos

© **2019, Forester de Santos**

ISBN: 9781070829517

All Rights Reserved

Kindle Edition

A bit about the Atom

The Atom is a microscopic magnet of matter or the element or even of existence.

When the Atom enters the vacuum of space from outside of space in the form of matter or element, the Atom is super-compressed by the vacuum of space and thus creating a microscopic big bang, all 118 of them!

After the microscopic big bang the magnetic field of the atom becomes a negative field due to the electron which now has separated from the atom's center or nucleus and takes position on what used to be the magnetic field of the atom...

Interestingly, even though the electron is speeding around the now electron field or ring, the electron also is composed of three main parts, but just as the atom is surrounded by a magnetic field or shield, the electron is also surrounded by an electron shield or shield or ring.

Tags: Atom, microscopic, magnet, matter, element, big bang, magnetic field, electron, proton, neutron, electron ring...

A bit about this Author Beloved

All writers must come sooner or later to the things that they want to truly write about and most come to write where the easy money is and most writers make a very big killing in becoming very rich and very famous by writing fiction and good for them!

But the great question is how much fantasy for the human race since the human race has being living in a fantasy since the race entered into consciousness or began to think and invent with words?

Well, that is a great or tall question that Forester de Santos as a writer has truly asked!

And so he has truly chosen to walk or to actually write on the road less taken or less written about!

And so he began to research and write about immortality and as the same goes, one becomes what one thinks or writes or even reads about the most!

Table of Contents

Prologue

The Atom is a microscopic magnet of matter or the element or even of existence.

When the Atom enters the vacuum of space from outside of space in the form of matter or element, the Atom is super-compressed by the vacuum of space and thus creating a microscopic big bang, all 118 of them!

After the microscopic big bang the magnetic field of the atom becomes a negative field due to the electron which now has separated from the atom's center or nucleus and takes position on what used to be the magnetic field of the atom...

Acknowledgement

I would like so very much to give thanks to my youngest of two sons for suggesting to me to write this short work about the Atom, the divisible atom.

Thanks Fred!

Dedication

The Atom is dedicated to all of those that really seek the truth so that the she frees them and they thus truly become for much more for they truly becoming for the truth herself through a higher or through a taller mental consciousness or life renewed...

Introduction

The Atom

[(+) (0) (-)]

The Atom is a microscopic magnet of matter or the element or even of existence, [(+) (0) (-)].

When the Atom enters the vacuum of space from outside of space in the form of matter or as an element, the Atom is super-compressed by the vacuum of space and thus creating a microscopic big bang, all 118 of them!

After the microscopic big bang the magnetic field of the atom becomes a negative field due to the electron which now has separated from the atom's center or nucleus and takes position on what used to be the magnetic field of the atom, [((-)) [(+) (0)] ((-))]...

Interestingly, even though the electron is speeding around the now electron field or ring, the electron also is composed of three main parts, but just as the atom is surrounded by a magnetic field or shield, the electron is also surrounded by an electron shield or shield or ring.

And so what we have here is negative going round negative going around negative, similar to our solar system in where the sun goes around the galaxy and the earth goes around the sun and the moon goes around the earth...

In other words, the electron also has its main parts of [((-) (0) (+))].

When all of the electrons wear out because of the friction between the electron and the nucleus, the protons begin to react causing an explosion and thus draining matter or the element from all energy.

In other words, when there is no longer any negative part or electrons, the positive part or the protons alone with the neutrons collapse on themselves causing an implosion and thus the end of matter or the element because of lack of energy.

Now, matter or the element is dust or waste in the vacuum of space or in the universe…

And the end of matter or the element is the end of light or illumination and thus the end of the universe…

To sum up, the Atom really is a microscopic magnet or diagram of matter or the element or even of existence.

The Atom is also the identifying number or unique tag of matter or the element, which is the purest form of matter.

There are only 118 elements in the universe or existence and every element has its unique atomic number, more like its proton number. No two elements have the same atomic or proton number.

In fact, no two elements have the same atomic weight, which is the weight of the protons, the neutrons and the electrons.

Also, the weight of the atom or the element is twice is atomic or proton number.

Interestingly, that as the electron becomes smaller because of the friction with the nucleus the nucleus becomes heavier.

The above phenomena can be compared with the earth and the moon. The earth gets heavier and the moon gets smaller.

At one point in time the earth will pull the moon to the earth provide the earth gravity or the earth's magnetic field stays strong enough to pull the moon to the earth, creating vast destruction to both the earth and moon...

Chapter One

[(+) (0) (-)]

What is, therefore, the atom?

The atom is a microscopic or is a physical miniature of matter, light or energy.

The atom is a microscopic or is a physical miniature of existence or of the universe or of physical creation.

The atom is a microscopic magnet inside a magnetic sphere or in a magnetic field composed of seven electron or seven magnetic rings.

The atom as microscopic matter is also composed or made up of three main physical parts. The three main physical parts are protons, neutrons and electrons or simply put, "+0-."

That is to say, A=+0-. In other words, the atom is plus zero minus, [(+) (0) (-)].

Atoms are also transitional or turn from positive to neutral to negative, [(+) (0) (-)].

Contrary to its name, the atom is physically divisible or is physically broken into three main physical parts.

And those three main physical parts can also be divided or can also be broken into many smaller physical and identical parts as the first three main physical parts.

The atom is also made of protons (+), neutrons (0) and electrons (-).

The seven electron or the seven magnetic rings of the atom are created by the atom's magnetic field or shield or the atom's magnetic sphere.

Actually, it is an atomic magnetic shield or an invisible magnetic or electron shell which are also called electron rings/

When the atom or when pure matter or when the element physically enters the vacuum of empty space, the atom is physically divided or physically broken or is physically separated from the atom's center or from the atom's neutrally charged nuclei or neutrally charged center, and the electrons occupy a specific physical part or a specific physical location or position of the magnetic field or the now electron ring.

What this actually means is that the atom went through a microscopic or a miniature physical expansion or the atom went through a microscopic or a miniature big bang!

Now, however, the first atomic number or the first element or element number one is the element Hydrogen

H is the symbol for the element Hydrogen.

Hydrogen has one proton or has one positively charged atomic or has one microscopic particle or has one atomic matter in the center or in the nucleus or nuclei of the atom.

The element Hydrogen also has one neutron or has one electrically neutral atomic matter particle in the nucleus.

The element Hydrogen also has one electron or has one negatively charged atomic or has one microscopic particle or has one atomic matter on the first electron or the first magnetic ring.

However, before Hydrogen enters the vacuum of space, Hydrogen really has two neutrons or two electrically neutral particles of pure matter.

When the element Hydrogen enters the vacuum of space, the element Hydrogen really losses one neutron as pure energy!

In fact, all matter or all elements or all atoms lost neutrons as pure energy in the vacuum of space.

Pure matter, the elements or the atoms have twice the number of neutrons before pure matter or the atom enters the vacuum of space.

For every one proton or for every one electron that enters the vacuum of space, one neutron is lost as pure energy in the form of heat. One neutron for each atom was lost. Pure matter, the elements or the atoms lost neutrons according to the atomic number.

For every proton or for every electron, one neutron is lost as pure energy in the vacuum of space.

The loss of neutrons as pure energy in the vacuum of space is 1, 2, 3, and 4, and so on until the very last atom or the very last element in the Periodic Table of the Elements.

The very last atom or the very last element is atom or is element number 118.

So, element or pure matter number 118 in the vacuum of space lost 118 neutrons as pure energy.

Helium, the second element on the Periodic Table of the Elements has two physical parts of atomic matter.

The symbol for Helium is He.

What this really means is that Helium has two atoms or has two atomic particles of pure matter.

That is to say, the two parts or particles in Helium means that Helium has two protons or has two positively charged pieces of pure matter.

Helium has also two neutrons or has two electrically neutral particles of pure matter in the center or in the nuclei.

For every one proton there is a neutron and there is also an electron. It is a simple diagram of a bar magnet, in where a proton is on one side, the neutron is on the middle, and the electron is on the other side, [(+) (0) (-)].

And because of the Big Bang the three main parts or the three main solid pieces of the atom or the magnet are separated from the center or separated from the nucleus.

The proton and the remaining neutron stay at the center or at the nucleus, but the electron takes position in the magnetic field or in the so-called electron or magnetic ring, [(+) (0)] (-).

Actually, the magnetic field or the magnetic ring is really a neutral or is an electrically neutral ring or field.

When the electrons are separated from the center or from the nucleus of the atom, the electron occupies the neutral ring making or turning the neutral ring into an electron ring or electron field or shield.

The electron transforms or charges the magnetic ring or shield into an electron ring or into a negatively charged ring,

To simply find the exact number of electrons or the exact number of negatively charged pure matter particles orbiting or going around the nucleus of an atom, one must count by two for the very first electron or for the very first magnetic ring and whatever number of electrons or whatever number of negatively charged pure matter particles remains goes to the second electron or to the second magnetic ring.

Helium, for example, has two electrons or has two negatively charged pure matter particles on the very first magnetic or the very first electron ring or the very first magnetic or electron orbit.

In reality, the Element Helium is an element made up or composed of two magnets united at the corners.

And being made up or being composed of two magnets, the Element Helium has two protons, two neutrons and two electrons, [(+)(0)(+)(-)(0)(-)].

Turning to the third element or the three whole or real numbers, count two electrons for the first electron or the first magnetic ring and count one electron for the second electron or the second magnetic ring.

And finally, turning to the fourth element or to the fourth atom, count two for the first electron ring, and there is a remainder of two for the second electron ring.

The fourth element is really four magnets united at a corner or united in links.

The same example or physical pattern as above follows on until the next neutral element or the next neutral atom, which is a gas.

Pure matter as gas or an element as gas is not electrically charged.

The next neutral element, which is Neon or element number ten, follows the very same and simple example or simple physical pattern of the first elements before Neon: count two for the first electron ring or the first magnetic ring, and count eight for the second electron ring or for the second magnetic ring.

The elements following Neon or following element or atom number ten follow the physical pattern of Helium, two electrons, and Neon, eight electrons.

The pattern for Neon is two for the first electron ring, eight for the second electron ring and whatever number remains goes to the third electron ring or goes to the third magnetic ring.

Element number eleven is counted two small cubes for the first electron ring, counted eight for the third electron ring or the third magnetic ring.

Element or atom number twelve is counted two for the first electron ring, counted eight for the second electron ring and counted two for the third electron ring or the third magnetic ring.

The very same physical example or the very same and simple physical pattern as above follows on until the next neutral element, and the very same and simple physical procedure as Helium and Neon must also follow.

That is to simply say, the next neutral element determines the number of electrons in electron rings or in magnetic rings.

The next physical break or the next physical division of the Periodic Table of the Elements, for example, determines the exact number of electrons that are going to the electron rings or to the magnetic rings.

It is already pre-determined even before matter or the original atom enters the universe or the vacuum of space.

As it can also be simply seen, the atom or the divisible atom is not only a simple microscopic diagram or a simple microscopic blue print of a bar magnet or a simple microscopic diagram of an electric motor or generator.

But also the divisible atom is a very simple microscopic diagram of the physical universe and of physical Existence or of physical reality or of physical creation and even the conscious mind and brain.

The divisible and the physical atom is actually a simple magnet inside another simple magnet, but instead of the two simple magnets physically attracting or physically uniting as two opposite charged magnets, the two atomic or the two microscopic magnets are kept physically apart by the magnetic sphere or by the magnetic field or shield or simply by a magnetic shell.

In a way, the magnetic sphere or the magnetic field of the divisible atom acts as a simple protective shield or acts as a simple protective sphere or act as a simple protective shell.

The simple magnetic sphere or the simple magnetic shell, therefore, keeps the divisible atom from physically collapsing or from further physical expansion or from further physical separation from the nucleus or from the center of the divisible atom.

That is the very simple reason that the element or that pure matter does not become physically much smaller or microscopic as the divisible atom.

In other words, the simple magnetic field or the simple magnetic sphere or the simple electron ring of the divisible atom or the element keeps matter from further physical expanding or from further physical collapsing or from physically becoming microscopic or from even physically imploding.

However, this does not mean that the magnetic sphere or that the electron ring cannot be acted upon from outside.

In contrast, interacting with the divisible and the physical atom from the outside creates mixtures of new elements or two or more elements united, thus making other elements or what is simply called in Modern Chemistry isotopes, compounds or mixtures.

This, however, does not mean that there will ever be more than 118 elements.

Chapter Two

The Small which makes the Great

[(+) (0) (-)]

Existence is infinite or without size. Existence runs or she expands toward all sides at the very same time and at the very same time existence also compresses toward all sides, that way also adding to her very self an infinite weight and an infinite size.

That is to say, existence adds to herself, she multiplies herself, she divides herself and existence also subtracts her very self or from her very self to be able to become forever for more and even for new.

Existence forever existed and existence forever will exist. Existence has no time even though time existence is.

Existence is also composed of three principal parts which all add to one and that one is the grandiose part which makes the difference in all of existence.

Existence consists of matter, that which is physical. Existence also consists of space or of vacuum, that which is lack of something or of matter.

And existence also consists of movement or of that which many call space-time, which is a movement as if in waves in space.

Thus in truth, existence is a simple magnet!

In other words, existence really is composed of +0-; in where (+) is equal to positive; in where (0) is equal to neutral; and in where (-) is equal to negative or the lack of.

In the scale of colors would be white, grey and black; in where white is light or matter; grey is neutral or space-time; and black is darkness or lack.

Now then, just as existence truly is composed of three parts and practically of parts opposite to the other parts, +0-, thus one also truly is composed of those very same opposite parts but in knowledge because one is knowledge.

And one can truly be positive knowledge or be neutral knowledge or even be negative knowledge.

But the greater the positive knowledge one has, less the neutral knowledge and less the negative knowledge or less the lack of negative knowledge.

In other words, the greater the knowledge of reality that one has, the greater the reality of one and less is the reality not known or less is emptiness or darkness.

Thus in truth, the more the knowledge of one, the more the abundance or the more is the life of one or more life makes sense to one.

Interestingly, that existence functions the same as one but existence knows it not but existence will never stop from being or will never stop from existing for all the eternity of eternity because she truly makes herself for much more!

That is in truth, even though all of existence is one, because of one existence truly is and for all eternity infinite!

And every time that existence is increased by one or is increased by more than one, the point of neutrality or the neutral is less as the same as the negative or the unknown point is less.

But making the things less does not make existence greater, but making the things or existence greater makes existence greater!

And just as existence makes the things greater for existence to be greater herself, oneself can be greater by recognizing that life is more or that in life herself there is more and that there is with all peace, the neutral, and that here is with all knowledge, one or more than one; and also that there is with all gladness and with all joy and that also there is with all abundance of life, life renewed…

Thus, through knowledge one renews for much more and because of one renewing for much more, one truly can continue for all of eternity as the very eternity…

Chapter Three

Creation and the Number

[(+) (0) (-)]

Creation truly is about giving name or naming to start or to begin and giving rename or renaming to continue without beginning again even though to give rename or renaming is to become as if forever new and as if never there were a start or a beginning and neither end...

All of existence truly is composed of numbers or be it positive numbers or be it neutral numbers or be it negative numbers or be it a combination of all the numbers at the very same time.

But there will always be a number that is greater than the other negative numbers until that positive number is converted into a neutral number and after that positive number is converted into a neutral number thus also it will be converted with time into a negative number.

In other words, existence really is equal to (+ 0 -), but existence does not return to negative but rather a part of her and that part of her is the universe or matter or light, which returns to or truly is converted from positive to neutral or to zero and much later it is converted to negative or into nothing.

24

The number zero is not negative or nothing. The number zero only is a neutral number which could be converted into a positive or negative number.

Existence as matter or as light or energy is truly composed of three parts which add to 118 percent, 118 times 3 because they truly are 118 positive parts, 118 neutral parts and 118 negative parts or (+ 0 -).

Now then, the neutral part or part zero is the universe and it is where it is added, it is subtracted, it is multiplied and it is divided at the very same time.

The positive part is the physical part of existence or the part from where comes out matter or light or the elements which truly are pure matter, but this part of existence appear to be minor than the other parts of existence.

The negative part of existence is the lack or is the vacuum of space or darkness which is composed of three parts, one is space or emptiness and the other two are space-time, which truly are created by the interaction of the magnetic field between the positive and the negative.

When matter from the physical side of existence enters into the vacuum of space, matter even though a single piece, matter enters with its 118 parts assimilating the 118 elements.

That is, if only one element enters into the vacuum of space that element could assimilate or even can come to be converted into the other 117 elements…

But in reality matter enters from the physical side of existence into the vacuum of space in 118 pieces into 118 parts of the vacuum of space or into 118 dimensions at the very same time.

That is to say, into the vacuum of space there enter 118 elements and every element has 118 parts and at the very same time there enter 118 elements into 118 dimensions.

But that does not remain like that, because with every element the 118 parts are multiplied.

In other words, element number one has 118 parts but element number two has 236 parts and element number three has 354 parts and this pattern continues on until the last element, which is element number118.

Also, the weight of the element is twice its number. That is to say, element number one has a weight of two and element number two has a weight of four and element number three has a weight of six. This pattern also continues on until element number 118.

Interestingly, that all the parts of an element add to the number of that element!

Thus, element number one has 118 parts and if we added those 118 parts thus we would get one.

That is to say, if we added 1+1+8 thus they would give us10 and if we added 1+0 thus we would get 1. In the case of element number two, if we added 2+3+6 thus we would get 11 and if we add 1+1 we would get 2.

This pattern continues on until element number 118. Element number 118 has 118 parts and every part also has 118 parts thus giving a sum of 13,924, also adding to one.

Now then, when element number one enters into the vacuum of space element number one enters with one proton, with one neutron and with one electron, (1+0-1). And also element number one enters with a weight of two.

But once in the vacuum of space, space comprises the element and the element is comprised causing the element to super heat and thus also causing the element to burst or explode.

In that burst or explosion the element loses a neutron or a third part and no longer the weight is of two but is less, such as 1.67.

Also the electron was separated and now is spinning around the proton or the nucleus of the element, (+) -.

In the case in where an element has a high number thus that element will neutrons in its nucleus, such as (+0+) -, -. Or (+0+0+) -, -, -.

This new transformation of the element, in where the electron is separated from the nucleus makes it possible for the element to be united to other elements and that way converting into a mixture of elements or isotopes...

Furthermore, before an element enters into the vacuum of space its three main parts have the same size or weight, ((+) (0) (-).

But once that element enters into the vacuum of space thus that elements is divided into two parts, the nucleus in where new is the proton (+) and the neutron (0), and the outside part in where now is found the electron (-) going around the nucleus, ((+) (0)) (-).

And while the nucleus keeps its size or weight, the electron loses its size or weight because of the interaction or friction which it has with the nucleus or with the center of the element.

The interaction or friction which the electron has with the nucleus or with the center of the element also causes the electron to last less or lasts less time in the vacuum of space.

Once the electron is fused or is exhausted, the element or the nucleus is converted into a neutral element or without energy even though the nucleus is still positive or with protons and neutrons.

But the outside part or the electric field of the element is now a neutral field or the electrons have been converted into neutrons because of lack energy.

In other words, the seven electron rings of the element or the atom now are neutral.

And just as the element functions thus that way also functions the number and existence herself, but the number or the symbol of the number is only an illustration of the numbers but truly does not show how is the number in existence or outside the vacuum of space in where there is no friction or movement even though there is a magnetic field.

Thus in truth, the number one or 1 outside the vacuum of space is represented by a cube. Now, the cube or the number or the element one is composed of three main parts and they are the positive part, the neutral part and the negative part.

But those parts also are composed of 118 other parts. That is to say, that the positive part also is composed of 118 parts and the neutral part is also composed of 118 parts as the same as the negative part which also is composed of 118 parts.

And when the cube or the number enters into the vacuum of space the cube or the number is compressed into a sphere or into a globe but still with its 118 parts.

Thus, the number one is composed of not only 100 percent but also of another 18 parts or of another 18 percent.

And if we added the parts thus we would have 1.

Now then, the number one or the symbol 1 as also is all of creation is a continuation because the start or the beginning is zero or a point or a neutral or an empty space.

But the number one or the symbol 1 also represents all of existence, the physical part, the neutral part and also the negative part.

Also the number one has the ability of converting itself into its 117 other parts also with their other 118 parts.

In other words, the number or element one also has the ability of being infinite because also it could renew into a greater number such as the number two.

And that makes it possible the other 117 parts which add to 9 and the number 9 is a symbol of renovation.

That first renovation extends the time of the number or of element one.

Thus, if we added the 117 parts which remain to one plus its other two parts, the neutral with its 118 parts, and the negative with its 118 parts, the sum would be of 353.

And if we added 353 thus it would give us 11 or eleven and if we added 11 we would get two, the possibility or the ability of the number or element one if it is renewed.

And once that the number or that element number one has renewed as two, thus the number or element one has become or will continue as double or for much more as two and as double the abundance.

And the very same step or process is with the number or with element number two. If we added all the parts that the number or the element two has, which are the double of one, thus it would give us four or 4.

That is to say, if we added all the parts of the number or element two thus we would get the double.

Chapter Four

The Understanding of Existence

[(+) (0) (-)]

The more understanding of existence or of reality one has of reality or of existence, therefore, the greater existence or reality will be to one as one will be to reality or to existence.

Reality or existence or the universe has to do with knowledge and acknowledgement, even unseen reality has to do with knowledge and acknowledgement even if negative knowledge or acknowledgement.

Space is an unseen reality or unseen knowledge or acknowledgement but space can be acknowledged but only when there is knowledge to acknowledge space with.

In other words, space can really be acknowledged through or because of matter or light. And matter as well as space can be divided into three major parts.

Space is not only emptiness or is lack but space is also dark and cold.

These three parts which make space are lack or are negative (-) knowledge or negative (-) acknowledgement.

Positive knowledge or acknowledgement, therefore, is matter and its three major parts, such as weight or mass, light and heat.

And matter only interacts with space while in the vacuum or emptiness of space.

Now then, existence is about knowledge and acknowledgement or confirmation and reconfirmation or on and off or 01.

When matter which is knowledge enters the vacuum of space, space begins to react with matter as a form of acknowledgement and the vacuum of space begins to compress matter until matter explodes and it begins to expand in the vacuum of space.

But the friction which matters has from space makes matter to last less in the vacuum of space.

Now, before matter entered the vacuum of space, space was in a state of tranquility or space was neutral or 0.

But when matter entered the vacuum of space, space acted as if negative due to the interaction with matter such as the compression of matter but when matter exploded and began to expand through the vacuum of space, space began to act as if positive.

But when matter begins to turn off due to lack of energy, space also begins to act negative.

And when black holes begin to appear due to the collapse of large stars, the vacuum of space begins to act even more negative.

And when there no longer is matter in the vacuum of space because the black holes vacuumed all the matter up, the vacuum of space has become completely negative due to the black holes now expanding throughout the vacuum of space in search of matter.

But just as matter ran out of energy, so will the black holes run out of matter and collapse or disperse and the vacuum of space will once again become neutral or return to 0 and once again become ready to receive matter or new knowledge or 1.

But this new beginning is as if the very first beginning because there will not be any memory that there ever was any other beginning or beginnings.

Also, the vacuum of space is dimensional and will receive matter or knowledge or 1 up 118 times.

That is to say, there is 118 beginnings which really begun at the same time or spontaneously and enter into 118 dimensions or space time.

But not only that, matter or knowledge or 1 enters the first dimension or space time as 118 percent and that 118 percent equals or adds back to one.

In the same manner, 2 equals 236 percent and that 236 percent equals or adds back to one and so forth with the other numbers up to 118. That is, 3 equals 354 percent and that 354 percent equals or adds back to one.

And finally, 118 or the last number equals 13924 percent, which is 118 times 118, and that 13924 percent equals or adds back to one.

The dimensions of time also add up as if they were one. And they increase in size from one to 118 times.

That is, the first is one but the second is twice as the first and the third is three times as the first but they still all add up as if they were just one dimension.

And if the speed of light was ever measured correctly the speed of light would be equal to one. The speed of light is about 186, 282 miles per second which adds up to 9 or point nine or 09.

This loss of 1 or of 01 is due to the compression of the vacuum of space. It is similar to water when it freezes. When water freezes water loses energy and weights from 1 to point 9.

Thus, existence or reality is one or existence or reality is knowledge or more like 01 which really means knowledge or acknowledgement.

For the conscious being, therefore, this information is very important because for this information the conscious being unknowingly struggles or contends for to be able to enter into higher or taller or greater conscious modes or be truly illuminated and through that illumination enter into a higher or taller or greater conscious existence and with this greater conscious existence or identity the conscious being can mode or transform existence or reality or his natural environment according to his personal will.

Of course, that forming or transformation really is done through his mouth and through the use of the spoken word or the zero plus one factor or language, more like binary because of the presentation or the knowledge and the acknowledgement.

Chapter Five

The plus Zero Negative Factor

[(+) (0) (-)]

Those that accept an idea or ideology blindly or without trying or testing that idea or ideology, thus they unknowingly become liars.

But nonetheless, liars they are and as liars they live and will lie to others to convince them to accept or believe and they will keep on lying until death herself and death herself will close their lying trap.

Now, when one does the movement or gets interested for knowledge, thus one truly becomes that knowledge.

When the universe or creation came to exist, the universe or creation came to exist because of knowledge.

The proof is in the numbers or in matter which is composed of elements and the elements in turn are really numbers, real numbers.

In other words, matter is knowledge because matter is composed of elements and the elements are composed of numbers and the numbers are composed of positive, neutral and negative states, thus the, [(+ 0 -)], plus zero negative factor or a magnet with its three opposites parts.

What this really means is that knowledge can become neutral or void or useless and then become negative or contra productive if at first nothing was done with that knowledge, such as using or converting knowledge into acknowledgement or a positive or useful respond.

In the same manner as above, the universe or the vacuum of space becomes positive when matter enters the universe or the vacuum of space.

But as matter begins to burn out, the universe or the vacuum of space begins to turn neutral until it turns negative, negative because of the black holes which now rule the universe or the vacuum of space which they now also suck up any remaining dust or matter to make space for another beginning.

But this new beginning is as if the very first beginning because there will not be any trace that there was ever a first beginning.

But the above matter would only be a theory or an idea or ideology if the conscious being, which is also real knowledge, did for real acknowledgement and the conscious being would have the power and the authority over matter to refresh matter and thus keep the universe always positive and refreshing in double abundance, all five portions of her!

Chapter Six

The Element

[(+) (0) (-)]

The universe is a neutral point or is a point zero which is converted into a positive point or into a one or more when light or matter enters into the universe.

The universe also is converted into a negative point when the grand majority of light turns off or the majority of matter no longer has energy...

Before light or matter enters into the universe the weight of the universe is cero and when light or matter enters into the universe thus the universe takes on weight.

Light or matter is composed of 118 Elements and the weight of each element is two times its atomic number, more like its positive number.

In the case of element number one, for example, its weight is of two and in the case of element number two its weight is of four and in the case of element number 118 its weight is of 236...

Curiously, that the totality of the 118 elements adds to one and that the totality of their weight adds to 2.

That is to say, that if we added from one to 118 thus the sum would be 7,021.

And if we added that sum thus it would be 10 and if we add that last sum thus it would be one.

That is to say, 118 is equal to one…

And if we added from two to 236 the sum would be 14, 042.

And if we added that sum thus it would be 11 and if we added that last sum thus it would be 2.

That is to say, 236 are equal to two…

Thus, light or matter enters into the vacuum of space or into the universe as one or as a unit which is composed of 118 pieces or elements and the weight of the element is two times the atomic number of the element.

Thus in truth, number one itself is composed of 100 percent plus 18!

That is to say, that the number one or even oneself is equal to 118 percent!

Also light or matter could enter into the vacuum of space or into the universe with only or as one element with its weight of two but that element would be able to convert into the other elements, even to the element 118 and its double weight of 236…

Element number one, for example, enters into the vacuum of space with its weight of two and it has 117 other possibilities of converting into the other 117 elements.

That is to say, element one is composed of 118 parts or pieces or the 118 percent and element number has 117 possibilities of converting into the other 117 elements according to the weight which element number one maintains.

In the same manner, element number two with its weight of four has 116 possibilities of converting into the other 116 elements or until the element number 118 with its weight of 236...

When light or an element enters into the vacuum of space, light or the element enters as if it were a piece of magnet or as if it were a bar magnet.

In the vacuum of space or in the universe the magnet or light or the element is super compressed not only until it takes the form of a sphere or round but also light or the element or the magnet is super compressed until it gets to a very high level of temperature.

And when the temperature gets to its highest level thus light or the magnet or the element super explodes causing the light or the magnet or the element to divide into two parts and the part with less weight, such as the negative part, takes position in the magnetic field and that magnetic field now is a negative field...

The superior or the positive or the heaviest part of light or of the magnet or of the element takes position in the center or in the nucleus.

Thus, now we have the negative part of the magnet going around the positive part when before they were united and the center or the nucleus was neutral...

And even though the element lost weight because of the explosion or because of bursting, the element continues the one for two.

That is to say, its weight continues of two although the element one now is 0.999 and its weight is double, 1.998…

Thus, now element number one was reduced to about 0.999 with its new weight of 1.998 but to element number one also remains a neutron or even more than one or a neutral part which can be converted or can be transformed into a positive part and that way not only adding the number of the element but also adding its weight even though it will have only one negative particle going around the center or the nucleus or the positive side…

But if one or light or the magnet or the element does not convert into the next number or into the next element thus it loses its energy and will only be a piece of dead matter in the vacuum of space and it will be removed one day by the black holes…

Thus, light or the element is the same as a negative particle, is the same as a neutral particle and is the same as a positive particle.

In a way, one is equal to a negative portion plus a neutral portion plus a positive portion which totality is of 0.999 after entering the vacuum of space.

But in the vacuum of space light or the element is positive even though the vacuum of space is neutral but reacts as if negative because of the vacuum.

And when the element increases its positive part by converting the neutral part into a positive part, the element cannot attract the negative part because of the vacuum of space because now the negative part becomes as if more or its weight increases because of the weight that it receives indirectly from the vacuum of space.

And if there were not positive attracting the negative thus the negative would expand through the vacuum of space and it would stop being negative and it would be dead matter...

Now, an element, in this case a star, which number is high as the same as its weight thus lasts or remains longer in the vacuum of space or in the universe.

But the element or the star becomes heavier while the energy or matter to continue on lasts or it begins to transform from positive to neutral and once neutral, the element or the star practically becomes negative when its excess or super weight attracts the electrons toward the center and thus causing an implosion in where the element or the star becomes a nova or a new star but without light or without energy and that way causing an enormous hole in the vacuum of space when before the element or the star occupied the vacuum of space as an element or as a star...

In other words, matter or the star in the vacuum of space changes from positive to neutral and then from neutral to negative.

Thus, [+ 0 -], in where the negative is a black hole or is a super vacuum cleaner in the vacuum of space or in the universe or in matter or the element.

This black hole practically eats or sucks all the matter around it to take all matter out from the vacuum of space or from the universe to make new space for new matter or for another beginning...

But as long as there are black holes in the vacuum of space or in the universe, thus the vacuum of space or the universe is negative and as long as the vacuum of space or the universe continues as or is negative thus it keeps being for something and not for neutral and from neutral or from zero to positive…

Thus, so that the universe becomes neutral or to zero and from neutral or from zero to positive thus all the black holes or super space vacuums must stop from functioning and once the black holes or the super space vacuums stop from functioning for lack of matter or for lack of energy thus the vacuum of space or the universe will return to neutral or to zero…

Now then, this new vacuum in space or in the universe is neutral and has zero energy but even so attracts new matter or new stars or attracts the new or the next beginning which is outside of the universe or outside the vacuum of space…

And once new matter or new stars enter into the vacuum of space or into the universe thus the vacuum or the universe will be positive or one or more…

Thus, the cycle or the model or the rhythm of (+ 0 -) continues until the end of all the times…

However, in the beginning or before the beginning, first there is the neutral or neutrality, (0), followed by positive or positivity, (+), which really is the beginning because of there being one or more; and then comes the negative or negativity, (-), for there lacking (0) or (+) or both.

Conclusion

The atom as also is the universe is finite or ends but existence is infinite or without size and has no end.

The atom or the universe ends or stops running but existence runs or she expands toward all sides at the very same time and speed and at the very same time and speed existence also compresses toward all sides, that way also adding to her very self an infinite weight as also an infinite size.

That is to say, existence adds to herself, she multiplies herself, she divides herself and existence also subtracts her very self or from her very self to be able to become forever for more and even for new or renewed but forever in perfect abundance and as if forever as if she was always new or renewed.

The universe begins as matter enters the vacuum of space, but existence forever existed and existence forever will exist. Existence has no time even though time existence is.

Existence is also composed of three principal or main parts which all add to one and that one is the grandiose part which makes the difference in all of existence.

Existence consists of matter, that which is physical. Existence also consists of space or of vacuum, that which is lack of something or lack of matter, energy or light.

And existence also consists of movement or of that which many call space-time, which is a movement as if in waves in space, more like a magnetic field.

Thus in truth, existence really is a simple magnet!

In other words, existence really is composed of [(+) (0) (-)]; in where (+) is equal to positive; in where (0) is equal to neutral; and in where (-) is equal to negative or the lack of.

In the scale of colors would be white, grey and black; in where white is light or matter; grey is neutral or space-time; and black is darkness or lack.

Now then, just as existence truly is composed of three parts and practically of parts opposite to the other parts, [(+) (0) (-)], thus one also truly is composed of those very same opposite parts but in knowledge because one is knowledge.

And one can truly be positive or real knowledge or be neutral knowledge or even be negative knowledge, just as false belief.

But the greater the positive knowledge one has, less the neutral knowledge and less the negative knowledge or less the lack of negative knowledge.

In other words, the greater the knowledge of reality that one has, the greater the reality of one and less is the reality not known or less is emptiness or darkness.

Thus in truth, the more the real knowledge of one, the more the abundance or the more is the life of one or more life makes sense to one.

Interestingly, that existence functions the same as one but existence knows it not but existence will never stop from being or will never stop from existing for all the eternity of eternity because she truly makes herself for much more!

That is in truth, even though all of existence is one, because of one existence truly is and for all eternity infinite!

And every time that existence is increased by one or is increased by more than one, the point of neutrality or the neutral is less as the same as the negative or the unknown point is less.

But making the things less does not make existence greater, but making the things or existence greater makes existence greater!

And just as existence makes the things greater for existence to be greater herself, oneself can be greater by recognizing that life is more or that in life herself there is more and that there is with all peace, the neutral, and that here is with all knowledge, one or more than one; and also that there is with all gladness and with all joy and that also there is with all abundance of life, life renewed...

Thus, through real knowledge one renews for much more and because of one renewing for much more, one truly can continue for all of eternity as the eternity herself...

Further Notes

Just as the atom is the greater part of matter or the element, thus our conscious mind is the greater part of the universe or of existence

And just as the electron transforms the atom's magnetic field into an electron field or ring so can we transform our conscious mind into a greater or higher conscious state by eliminating our dead beliefs or our dead faith which will take us dead to the grave if we do not remove our dead beliefs or our dead faith and that includes blind faith from our conscious mind...

Seek the Truth

Seek always of the truth or seek the truth and when she let herself be found from you, you truly will become taller, bright and will even have more confidence in yourself.

The very curious thing about the truth is that she begins with one and if one truly does the movement to find her, thus she will come to one or she will be given to you by he that made or that presented her for the truth.

A Higher Mental Consciousness

The path of less resistance or without contention not only leads to laziness or to vanity or to nothing but also the path of less resistance or without contention leads to death.

And death ends any possibility to any other possibilities, such as a higher mental consciousness or illumination with real gladness and with real joy, in double abundance all five portions, and also with the power and with the authority of the heavens here on the earth.

The lonely road to a higher or taller consciousness

A short auto bio

(1)

A funny thing happened to me on the lonely, narrow and unbeaten road to rebirth or to life renewed, which really is to a higher or to a taller consciousness or an expansion of mental consciousness.

At first, and this was even before I took that first step on the lonely, narrow and un beaten road or path to a higher or a taller consciousness, I wanted to be known.

I wanted to attract countless friends. I also wanted to be popular and perhaps get free lunches or dinners.

But once I took that first step to a higher or taller consciousness, my so called life became lonely, especially when I turned to my so called friends and family.

They marked me, but not before laughing or making jokes.

Some of them even recommended me to professional help.

And so I discovered that one cannot share one's findings with just anyone.

But the irony was that there was not anyone for being something new or the start of a new life or life renewed, what some call rebirth or born again.

Most people like to hear of things that they already know about and share or speak of thing that are common or simple, such as the news of the day, the weather or social or family events as long as they do not require too much conscious effort from their part.

And so I entered alone into the lonely, narrow and unbeaten road or path to a higher or taller mental consciousness or rebirth…

(2)

The reason that I walk alone

After trying to share my life improving findings with the people that I liked and even loved, I decided to keep quiet and walk alone but not lonely or feeling alone or even sad on the unbeaten road or path to a higher or taller mental consciousness.

Interestingly, that when I first started my grandiose search and research, I began to feel gladness and joy and even also I felt abundance or a pleasant presence…

At that moment I realized that there was more in the world and more to life and so I began to believe or to have faith. I also realized that I needed to acquire or achieve that higher or taller mental consciousness or intelligence.

At first I did not know that I had entered into a higher or taller mental consciousness, but when I began to think of greater or taller things in the world and to see people as they really are, I knew then that I had entered into a higher or taller mental consciousness or state…

I also began to study and read as many books as I could on the subject of a higher or taller mental consciousness, but the close that I ever got was that after life or after death was the answer or the truth.

But I totally rejected that notion or idea or rumor that death was the final frontier…and that more was beyond the grave or after death…

I had read so many books that I realized that most of them were inter connected to each other and that their only purpose was to make someone rich and famous and also to waste one's limited time…

But the books were not the only things wasting one's limited time. Unfortunately, everything else was, from smart cell phones and computers and televisions and even so called news to even friends and family members!

However, there is no problem with the above matters if one draws a limit or a line and one sets oneself a purpose or a goal to go beyond one's present mental capability.

And that was what I did. I put a limit to everything that did not take me to my purpose or goal, which was to the next consciousness or further into the unbeaten or untraveled road or path.

I even began to physically exercise and take vitamins and to eat as healthy as I could. I lost a lot of weight, mostly fat around by belly.

My waist size went down from 40 inches to 36 inches. My bad cholesterol went down from 300 to 100. My sugar level in my blood before breakfast was 80!

People asked me if I was in love. I answered that there was nothing wrong with that, even though my only love was with my soon to be life renewed, which is really a higher or a taller mental consciousness.

And some others called me youth or young guy and smiled at me.

They even told me that I may get a speeding ticket for running so fast!

To be continued...

Other Related Works

The Universe

The Universe or the vacuum of space really is a neutral point or point zero or a starting point and the universe becomes positive or turns into a one or a beginning when matter enters the universe or the vacuum of space from the outside of the universe or the vacuum of space.

The universe becomes negative when the largest of stars implode into giants black holes.

It is through the giant black holes that the universe or that the vacuum of space is cleared or cleaned from any matter or stars or planets which are still left in the universe or in the vacuum of space and thus returning the universe or the vacuum of space into a neutral point when all of the giants holes are dispersed or turned off because of lack of energy or lack matter in the universe or in the vacuum of space.

Tags: universe, vacuum of space, matter, energy, light, black hole, positive, negative, neutral, beginning, end, and implosion, planets

The Human Brain

The human brain is an exact copy or a duplicate of existence or even God who is a greater or taller consciousness and he that understands existence or God thus understands the human brain and he that understands the human brain understands existence and he will be able to persuade existence or the brain or even God to do his will.

But that persuasion of existence or the brain or even God has everything to do with mental consciousness and he that knows the power of mental consciousness will make existence or his brain his true reality.

In other words, he will be able to really transform existence or transform matter according to his will. However, that requires a clear conscious mind or a conscious mind without false beliefs or a conscious mind with true knowledge

Tags: The Human Brain, cerebrum, mental consciousness, point of view. Police detective, researcher, writer, educated guess, conscious mind, conscious thought, global economy, Stone Age, nightmare, dream, black holes, after death, Paradise

The Black Hole

The Black Hole is a giant hole or a round rupture or tear or breakage in the fabric or the foreskin skin which really is the vacuum of space.

But the hole or the tear or the breakage in the vacuum of space moves along the vacuum of space sucking up or removing or destroying all matter, including all the stars and planets and even all forms of life, to clear the universe or the vacuum of space for new matter or for another beginning which will appear as if the very first beginning because there will not be any trace that there was ever a first beginning...

Tags: black hole, vacuum of space, stars, planets, beginning, universe, matter, breakage, foreskin, cosmos...

The Origin of the Universe, the complete verse

The Origin of the Universe, the complete version, is about the true and the only origin of the universe.

The tall title says it all.

Those conscious beings that do not know reality or do not know the truth thus they will not be able to live in reality or in the truth nor will they live as reality or as the truth they them very selves and they live as if they do not live.

And in truth, they will die because of the lack of reality or because of the lack of the truth, that if recognizing reality or the truth will give them true life and in true abundance of peace, knowledge, joy and gladness.

Tags: Origin, universe, cosmos, existence, matter, space, black holes, space-time, atom, element, table of elements, periodic table of the elements, rings, Seven

My Inspired Words for Royal Life

My Inspired Words for Royal Life are some essays or verses, some short and others long, which explain not only the importance of knowledge and of acknowledgement in the conscious being but also explain how existence truly functions or the grandiose purpose of her or for her and how to truly achieve royal life or the royal salvation by the act of oneself...

Thus in truth, fortunate is he that has the grandiose opportunity of being able to read **My Inspired Words for Royal Life** because he will be amazed and truly will be a better conscious being or will be taller being in identity...

Tags: royal life, royal peace, royal knowledge, royal salvation, the kingdom of the heavens, immortality, royal faith, new heavens, new earth, servant beloved of God, neutral point, purpose, foresters, creation, procreation, adoption, repentance, 01

The Origin of the Universe

A Short Version

The origin of the universe, finally the last chapter!

All those conscious living beings living today who truly want to know the true origin of the universe, the true origin of life, the true origin of thought and the true origin of the beginning and the end, this is the eBook or printed book to read!

Now then, everything has an answer and the answer starts with one or as one when one questions to know and thus to have knowledge to knowledge thus one becomes, becomes because the purpose of knowledge is thus one as one is thus knowledge...

Thus, truly blessed is that man that has true knowledge of how existence or the universe truly functions, because that man also knows how he himself thus functions.

And he also will do as existence herself to continue on existing or living on but existing or living on as if forever new and in complete, which is perfect, abundance, all five portions of her.

The Real Origin of the Universe

The Real Origin of the Universe is about the true and thus the only origin of the universe.

The tall title says it all. Those conscious beings that do not know reality or do not know the truth thus they will not be able to live in reality or in the truth nor will they live as reality or as the truth they them very selves and they live as if they do not live.

And in truth, they will die because of the lack of reality or because of the lack of the truth, that if recognizing reality or the truth will give them true life and in true abundance of peace, and real knowledge, joy and gladness.

Who am I really?

(1)

My pen name as a writer is Forester de Santos and I am truly on a very grandiose crusade of rebirth alive or to be born again into royal life or life renewed with complete gladness and with complete joy and also with complete abundance of God, which is a higher mental consciousness, but God as much more than God and as much more than Creator.

Now then, one who truly is on a very grandiose crusade cannot follow another or cannot let himself be surrounded by his beloved ones or his fans because he cannot cross over them or he cannot cross over because of them being in the way or because of them blocking the path which is but which cannot be seen until rebirth or until one is born again.

I do not ask to be followed, not because I will not lead, but because I will not look back but I will look to my right and to my left to see who walks with me.

But those that truly decide to follow me will become as me and as me will truly receive or gather true knowledge because my struggle or contention or my very grandiose crusade of rebirth is true, so true in fact that I have become a much better person because of the true faith which I have come to receive through my search and research for the truth.

And because I have come to have true faith or faith of God, thus I use my true faith as a shield to repel or to reject other beliefs or good sounding lies!

Therefore, to rebirth alive or to be born again while still living here on the very earth which will be as in the very heavens through rebirth!

Now then, God could be the Master Creator, the brain, the sub-conscious mind, the conscious mind surpassed into a taller consciousness or even much more, but all of these forms or reforms of God or of life forever will be every time higher in consciousness and so that man could enter or overcome a higher conscious thus man has to present himself for more or as more to what is higher than man so that which is higher than man, God, allows man to go up to the higher or to the next mode of thinking or of mental consciousness...

$$(((((((0+1)))))))$$

If you truly enjoyed this simple and humble work, please leave a comment according to your good pleasure and give also a rating but also according to your good pleasure.

Thanks so very much for your time and best of wishes, Forester de Santos.

Thanks for reading my work!

0+1 = peace and knowledge to all mankind!

Para sus notas personales

Para sus notas personanles

www.ingramcontent.com/pod-product-compliance
Lightning Source LLC
Chambersburg PA
CBHW020708180526
45163CB00008B/2995